AF322679

When the World Learned to Count

Contents

Author's Note

For Kindle—

A beautiful, intelligent young woman whose belief in children is louder than the world's doubts.

You remind us that learning is not just about numbers, but about curiosity, courage, and care.

You stand where it matters—
with children,
for children,
guiding them not just to count the world,
but to understand it.

This book carries a small piece of that spirit.

May every child who opens these pages
feel seen, encouraged,
and inspired to wonder a little more.

Chapter 1: Before Numbers Had Names

Did you know people were counting even before number had names? "Before numbers had names, people still counted steps, days, stars, and the hands they held."

Leaves whispered, "one,"

Raindrops answered, "two,"

And the stars blinked, "three."

People counted what mattered to them.

In a small village under a cotton tree, a girl named **Omi** looked up and asked, "Who taught the world to count?"

A golden firefly landed on her thumb. Its light flickered like a secret waiting to be told. The firefly chanted:

"Raindrops taps on window glass, —one, two three."

"The wind whispered totals through the grass, — four five six."

"And every heartbeat softly said— seven, eight, nine — please be mine."

Then the firefly said "I can show you how the world learned to count," touch my wings and hold tight — we'll visit the counters of time." And off they flew, into the history of how the world learned to count.

Chapter 2: Africa Counted with Rhythm

From above Omi could see a fire with people sitting around it in a circle. The firefly took her to the site so she could get a close look. The firefly whispered. "Before symbols, there were songs. Before paper, there was rhythm." Around a fire, drummers played stories, while a griot spoke in rhythm. Each beat told a number:

- One beat — "the first light."
- Two beats — "the pair that made the world."
- Three beats — "the harmony of all things."

"We counted what we could feel," said the griot. "Sound is a kind of number too." Omi clapped, feeling the numbers in her bones. The firefly pulsed to the rhythm. The drummers sat in a circle, their hands alive.

Boom — one. Boom-boom — two. Boom-boom-boom — three. The griot sang: "Before we drew numbers, we *sounded* them. Before paper, there was memory." He tapped his chest. "Here is where we kept them safe."

The Beads and the Drums

Beads and knots could be used for counting:

One drumbeat for home. Two for family. Three for the ancestors who listen.

Omi joined the rhythm — clap, clap, clap.
She felt the count travel from her palms to her heart.

The firefly pulsed with the beat, a tiny metronome of light. "See?" said the griot. "Even the air keeps time."

The fire crackled softly. "Listen," said the griot, his drum speaking with warm hands —
boom... boom-boom... boom.

"Each beat means something," he told Omi. "One for the sun, two for the moon, three for the stars that never sleep." Children laughed as they clapped along. A grandmother beside them tied a rope with gentle fingers. "When we had no paper," she said, "we tied our stories. Each knot kept a number safe — one cow, two baskets, three brave sons."

Omi strung colorful beads beside her — red for the sunrise, blue for the river, green for the harvest. "We count what we love," whispered the griot.

The firefly danced in rhythm, glowing once for each beat.

Clap one time — the day's begun
Clap two times — here comes the sun
Clap three times — we laugh and play
Clap four times — we count the day

Stomp your feet — five, six, seven
Spin around — eight, nine, ten

Clap again —

Now you see
Counting lives inside of me.

Its time to go Omi.

"Touch my wings," said the firefly.
And off they went.

The Nile and the Lotus
Egyptian Hieroglyphs
1   to   und   100
10
100

Chapter 3: Egypt Counted the River

Next, Omi saw a great river, its tides powerful as they moved. They stopped by the banks of the river Nile on the same continent. She asked a young boy nearby , "do you count and why do you count?" He replied.

"Every year, the river changed. Sometimes it was quiet and calm. Sometimes it rose and spread like a shining blanket across the land. My people watched carefully writing down every sign."

Omi sat by the Nile with the boy who knew its moods. They listened to the water together. "When the river comes," the boy said, "it brings life. "When it leaves, it leaves work." Omi watched as the water climbed the banks and then slowly slipped back home again.

"How do you know when to plant?" Omi asked. "How do you know when to wait?" The boy smiled and drew in the sand. He made one small mark. "This is one season," he said. He added another. "This is the next." Soon, the marks became pictures.

One stroke for one —

A heel bone for ten —

A coil for a hundred —

A lotus for a thousand —

And a god with open arms for a million —

"We count the river, the boy said, because the river counts us. We counted the days the water stayed. We counted the days it fell."

They counted the months between floods. With counting, they knew, when to plant the seeds, when to store the grain, and when to prepare for the next rise. People counted what mattered to them.

Omi dipped her fingers into the sand and copied the pictures. "These are numbers," she said softly. The boy nodded. "They are memories," he answered. "Of water. Of time. Of trust."

The firefly hovered above the river, glowing like a small sun. Omi looked at the Nile and understood.

Some people counted beads. Some people counted beats. And Egypt counted the river — so the land could live again.

The firefly led Omi to a quiet place where the ground was soft, and the stars were busy. A priest pressed marks into clay. Some were close together. Some were far apart. Omi leaned in. "They look the same," she said." How do you know what they mean?"

The firefly hovered between two marks and stopped. "Because of the space," it whispered. The man pointed to the clay. "When marks stand together," he said, "they tell one story. When they stand apart, they tell another."

Omi smiled. "So even nothing can help count," she said. "Because without space," the priest said, "everything would become one long sound."

Omi nodded slowly. "So even what isn't written... still matters." Africa counts in many ways, by color, by sound, by symbol, by sky."

The firefly glowed brighter. "Especially nothing," it replied.

"Touch my wings," said the firefly.
And off they went.

一 二 三 四
丿 五 六 七
十 八 九 十

Chapter 4: China Counted with Lines

The firefly carried Omi east and land her in a quiet garden where bamboo swayed and ink breathed. An old scholar dipped his brush where ink met silence and painted:

一 二 三 四 五 六 七 八 九 十

"One, two, three," he said, smiling. "Each stroke must balance heaven and earth." "Too many strokes and you lose the wind."

 He showed her 四, a square for four corners.

Then 五, the five elements — wood, fire, earth, metal, water.

Then he showed more:

四 (four corners),

五 (five elements),

六 (six directions).

"Numbers remind us how the world stays steady," he said. "If one side is heavy, the other must give." "Numbers are not only for counting things," he said.

"They are for *keeping balance*." Omi painted beside him. Her strokes danced crookedly at first, then steadied. "They look like the way the wind breathes." "They feel alive," she said.

He nodded. "Because they are. In every line, the world breathes once."

And with each breath we are counting our world around us. Repeat after me.

One is a line — just stand up tall
Two is a step — don't fall at all
Three is a rhythm — tap, tap, tap
Four makes a box — snap, snap, snap

Five bends softly — watch it move
Six finds balance — find your groove
Seven leans, eight opens wide
Nine curls in — ten stands with pride

Line by line, the numbers grow
Count them once… now say it slow

Omi chanted the lines in rhythm, until she had them memorized.

"Touch my wings," said the firefly.
And off they went.

Chapter 5: The Maya Counted Time

Next came a jungle alive with drums and parrots crowing, where women carved dots and bars into smooth stone.

"Each dot is one," said a carver. She pressed her finger: •

"Each bar is five," she said, carving one bold line: —

Omi counted:
One dot = 1, two = 2, three = 3,
one bar = 5,
one bar and two dots = 7.

Then the woman pointed to a carved shell. "What is that shell?" Omi asked, pointing.

"That," the woman whispered, "is **zero** — our breath and seed of time. "Without it, our calendars would never work."

Omi touched the shell's spiral. "It's empty but endless," she said. "Exactly," said the woman. "Zero lets us begin again — like the moon after darkness."

Some people didn't just leave space they gave it a name.
Omi spun in a circle, whispering 'zero' until she laughed
and fell still.

"Touch my wings," said the firefly.
And off they went.

I
IV
IV
II
VI
III
VII
IX
XIII

Chapter 6: Rome Counted with Hands

They landed where marble shone bright and white, and builders sang in rhythm with their chisels. "Watch," said a man shaping stone.

He carved a straight line: **I** "One pillar," he said.

He carved another: **II** "Two pillars."

Then three, then four — and when he reached five, he smiled. He spread his fingers wide.

"Five fingers — that's **V!**" Omi laughed.

"Your hand looks like it!" "Yes," said the builder. "We built our numbers the way we build walls — from what we can hold."

He showed her:
I (1), II (2), III (3), IV (4), V (5),
VI (6), VII (7), VIII (8), IX (9), and then—
two hands crossed — **X** for ten.

"Why X?" Omi asked. "Because two hands make ten," the builder smiled, showing open palms. "We count with what we carry.

"Two fives make an X," Omi whispered. "Ten fingers, ten stones, ten ways to begin again." The builder nodded.

"We Romans counted to build — so our numbers stand tall, like soldiers who never forget their place." "Numbers were our tools," he said. Straight as pillars.

Honest as stone."

1, 2, 3 — just me, myself, and me
4, 5, 6 — we're all in the mix
7, 8, 9 — say it one more time
10 — let's do it again!

"Touch my wings," said the firefly.
And off they went.

Zero
The Nothing
That Changed
Everything

Chapter 7: The Arabic World Shared the Numbers

Over dunes of gold and wind, they came to a scholar drawing in sand by starlight. In the desert, the stars looked like lanterns on paper. "Each number," he said, "was born from an *angle* — the meeting place of thought."

He pressed his stick and drew:

> 1 has **one** angle.

> 2 has **two** angles.

> 3 has **three**, and so on up to 9.

Each with its own sharp corners of reason.

Omi leaned close, tracing each line. "So, the shapes mean something?"

"Each number," he said, "was once made of *angles* — the language of geometry."

"So that's why they look sharp!"

"Yes," said the scholar. "They were made to be read by the eye and measured by the mind. Numbers were once geometry — truth written small."

Then she saw it — a circle drawn in the sand. "What is that?" she asked. "That," he said softly, "is **zero** — the nothing that changed everything."

"Nothing?" Omi frowned "It looks like something."

The scholar smiled. "Exactly. Zero is the space between all somethings — the silence that makes music possible."

Omi leaned close. "That," she smiled, "is *zero.*" The nothing that changed everything.

The firefly flew in circles making many zero.

"Touch my wings," said the firefly.
And off they went.

33

Chapter 8: The Gift of Zero

When the journeys were done, Omi looked at the firefly. It spun in the air and drew a shining circle. "Zero," it whispered, "is the space between every number. It is the pause that gives meaning to sound."

Omi smiled.
"So all counting begins... with nothing?"

"Yes," said the firefly.
"And that nothing became everything."

Remember to count your heartbeat, to count the stars, to count the ebb and flow of the mighty rivers and remember to leave space for zero.

Count What Matters

Count your steps
Count your days

Count the stars
Count your ways

Count your laughter
Count your tears
Count your stories
Through the years

Some count fast
Some count slow
Some count things
We'll never know

But every voice
In every land
Counts what matters
Close at hand

Say it once — now say it true:
The world learned to count... through you

Omi was
home again.

She counted her steps.
She counted the stars.
She counted the hands
that held hers.

And this time...
she knew why.

Chapter 9: For Curious Minds (Grown-Ups Welcome)

Here's what Omi remembered:

Roman Numbers

- I = 1, V = 5, X = 10, L = 50, C = 100, D = 500, M = 1000
- X came from two hands crossed — ten fingers!
- Romans didn't use zero. That's why their math stopped short.

Chinese Numbers

- 一 = 1, 二 = 2, 三 = 3, 四 = 4, 五 = 5, 六 = 6, 七 = 7, 八 = 8, 九 = 9, 十 = 10
- Each stroke shows balance and simplicity.
- Counting rods helped them do early calculations.

Arabic Numbers

- 1–9 were once drawn from angles.
- Zero (0) came from ancient India and traveled through Arabic scholars to the world.
- Without zero, there would be no computers, no place values, no infinity.

Mayan Numbers

- Dots = 1, Bars = 5, Shell = 0
- They used base-20 (not base-10 like us) — perfect for tracking moons and seasons.

African Counting

- Rhythmic and oral — numbers kept in songs, drumbeats, and dances.
- Memory and sound became early mathematics.

Some numbers stand straight; some curve or hum —but all of them whisper the same thing:

Count what matters. And never forget where counting came from.

People counted what mattered to them — and then they
shared it.

Remember to count your heartbeat.
Count the stars.
Count the rise and fall of rivers—
and always leave space for zero.

Where the World Learned to Count
West Africa –
Beads & Knots
Egypt –
Hieroglyphs
& Nile
Arabic Numbers
& Zero
1 2 3 4 5 8 9
China –
Chinese
Characters
中文
和三
Central
America
Dots & Bars
N
E
S
W

9 798999 284827